OEE for Operators

SHOPFLOOR SERIES

OEE for Operators
Overall Equipment Effectiveness

CREATED BY

The Productivity
Development Team

CRC Press
Taylor & Francis Group
Boca Raton London New York

CRC Press is an imprint of the
Taylor & Francis Group, an **informa** business

CRC Press
Taylor & Francis Group
6000 Broken Sound Parkway NW, Suite 300
Boca Raton, FL 33487-2742

© 1990 by Taylor & Francis Group, LLC
CRC Press is an imprint of Taylor & Francis Group
Originally published by Productivity Press
Cover by Mark Weinstein, Cover illustration by Gary Ragaglia,
Graphics by Guy Boster, Lee Smith and Hannah Bonner.
Printed and bound by Malloy Lithographing

No claim to original U.S. Government works
Printed in the United States of America on acid-free paper
International Standard Book Number-13: 978-1-56327-221-9 (Softcover)
20 19 18 17 16 15 14 13

This book contains information obtained from authentic and highly regarded sources. Reasonable efforts have been made to publish reliable data and information, but the author and publisher cannot assume responsibility for the validity of all materials or the consequences of their use. The authors and publishers have attempted to trace the copyright holders of all material reproduced in this publication and apologize to copyright holders if permission to publish in this form has not been obtained. If any copyright material has not been acknowledged please write and let us know so we may rectify it in any future reprint.

Except as permitted by U.S Copyright Law, no part of this book may be reprinted, reproduced, transmitted, or utilized in any form by any electronic, mechanical, or other means, now known or hereafter invented, including photocopying, microfilming, and recording, or in any information storage or retrieval system, without written permission from the publishers.

For permission to photocopy or use material electronically from this work, please access www.copyright.com (http://www.copyright.com/) or contact the Copyright Clearance Center, Inc. (CCC) 222 Rosewood Drive, Danvers, MA 01923, 978-750-8400. CCC is a not-for-profit organization that provides licenses and registration for a variety of users. For organizations that have been granted a photocopy license by the CCC, a separate system of payment has been arranged.

Trademark Notice: Product or corporate names may be trademarks or registered trademarks, and are used only for identification and explanation without intent to infringe.

Library of Congress Cataloging-in-Publication Data

OEE for operators : overall equipment effectiveness / created by the Productivity Development Team.
    p. cm.
  Includes bibliographical references.
  ISBN 978-1-56327-221-9
  1. Total productive maintenance. 2. Industrial equipment. Productivity Development Team (Productivity Press).
TS192.032   1999
658.2'7—dc21   99-34532

Visit the Taylor & Francis Web site at
http://www.taylorandfrancis.com

and the CRC Press Web site at
http://www.crcpress.com

Contents

Publisher's Message   xi
Getting Started   xv

The Purpose of This Book   xv
What This Book Is Based On   xv
Two Ways to Use This Book   xvi
How to Get the Most Out of Your Reading   xvii
Overview of the Contents   xix

Chapter 1. About TPM and OEE   1

Chapter Overview   1
What Is TPM?   2
What Is OEE and Why Is It Important?   4
    Quantity Over Time Is Only Part of OEE   4
    Effectiveness Focuses on the Equipment, Not the Person   5
    The Purpose of Measurement Is Improvement   5
The Role of the Shopfloor Team in Using OEE   6
In Conclusion   7
    Summary   7
    Reflections   8

Chapter 2. Understanding Equipment-Related Losses — 9

Chapter Overview — 9
Losses Reduce Overall Equipment Effectiveness — 10
Visualizing OEE and the Losses — 12
    Availability — 12
    Performance — 13
    Quality — 13
Availability: Downtime Losses — 14
    Failures and Repairs — 14
    Setup Time — 14
    Other Losses to Availability — 15
Performance: Speed Losses — 18
    Reduced Operating Speed — 18
    Minor Stoppages — 18
Quality: Defect Losses — 20
    Scrap and Rework — 20
    Startup and Reduced Yield — 20
In Conclusion — 22
    Summary — 22
    Reflections — 23

Chapter 3. Measuring OEE — 25

Chapter Overview — 25
Closing the Feedback Loop — 26
Collecting OEE Data — 26
    Defining What to Measure — 26
    Making Data Collection Simple — 29
Processing OEE Data — 33
    The OEE Calculation — 33
    Storing OEE Data — 34
Reporting OEE Results — 35
In Conclusion — 37
    Summary — 37
    Reflections — 38

Chapter 4. Improving OEE — 39

Chapter Overview — 39
5 Why Analysis — 41
Autonomous Maintenance — 42
Focused Equipment and Process Improvement — 44
Quick Changeover — 46
    Stage 1: Separate Internal and External Setup — 46
    Stage 2: Convert Internal Setup to External Setup — 47
    Stage 3: Streamline All Aspects of Setup — 47
ZQC (Mistake-Proofing) — 48
    Poka-Yoke Systems — 49
P-M Analysis — 50
In Conclusion — 54
    Summary — 54
    Reflections — 56

Chapter 5. Reflections and Conclusions — 57

Chapter Overview — 57
Reflecting on What You've Learned — 58
Opportunities for Further Learning — 59
Conclusions — 59
Additional Resources on TPM, OEE, and
  Equipment-Related Losses — 60
    Training and Consulting — 60
    Packaged Education and Support — 60
    Conferences and Public Events — 62
    Newsletter — 63
    Website — 63

About the Productivity Development Team — 65

Publisher's Message

Smoothly operating equipment is critical for manufacturing today. Most processes use machines to add the value customers pay for. In an environment that is more competitive than ever, factory machines have to work dependably to supply products when the customer needs them. Yet factories everywhere are plagued with machine problems of one type or another. The companies that are pulling ahead in the production race are those that understand their equipment problems and take steps to eliminate them. The key to this understanding is overall equipment effectiveness.

Overall equipment effectiveness (OEE) is a measure that shows how well the equipment is running. It indicates not just how many products the machine is turning out, but how much of the time it is actually working—and what percentage of the output is good quality. Because it reflects these three important things, OEE is an important indicator of the health of the equipment.

The condition of the equipment isn't just a maintenance issue anymore. In Total Productive Maintenance (TPM) approaches, equipment operators help prevent equipment problems through their knowledge and familiarity with the machines. Operators also monitor the machine conditions used for calculating OEE. This book is intended to share basic learning that will help you participate effectively as your company applies OEE and begins to reduce equipment-related losses.

Chapter 1 lays a foundation with basic definitions related to Total Productive Maintenance and OEE. You will learn why it is important to track effectiveness rather than efficiency. Chapter 2 introduces the three elements of OEE and their connection to key types of equipment-related losses—problems and wastes that reduce a machine's effectiveness. This is a basic framework that can be adapted to measure and begin to improve equipment problems in many different industries.*

---

*The OEE calculation and loss framework used in this book relates most directly to discrete parts manufacturers, rather than process industries, which face slightly different issues. For more on measurement in process industries, see Suzuki, ed., *TPM in Process Industries* (Productivity, 1994).

PUBLISHER'S MESSAGE

Chapter 3 offers a step-by-step overview of the process of doing the OEE calculation.* One basic aspect is shopfloor involvement. It's important for data to be collected on the shop floor and turned into information for use on the shop floor—not confined to an office or information department. This chapter also describes how to define what to measure and how to collect and process OEE data. It gives examples of different information displays that OEE data can generate (computer software is helpful for this).

Chapter 4 talks about how to respond to OEE information to fix the problems. It introduces the 5 Why method, autonomous maintenance, focused equipment improvement, quick changeover, mistake-proofing, and P-M analysis. Chapter 5 helps you review your learning and suggests additional resources for exploring key topics.

It is important to remember as you read that this material is a general orientation to a complex topic. Application and mastery of overall equipment effectiveness often requires a deeper understanding of the production mechanism. The process of using OEE is best supported by experienced consultants and trainers who can help you tailor it to your company's specific situation and address issues that may come up.

This book incorporates a number of features that will help you get the most from it. Each chapter begins with an overview of the contents. The book uses many illustrations to share information and examples in a visual way. Icon symbols in the margin flag key points to remember in each section. And "Take Five" questions built into the text provide a framework for applying what you've learned to your own situation.

One of the most effective ways to use this book is to read and discuss it with other employees in group learning sessions. We have deliberately planned the book so that it can be used this way, with chunks of information that can be covered in a series of short sessions. Each chapter includes reflection questions to stimulate group discussion.

---

*Some traditional approaches to OEE use a two-part formula for calculating performance that uses cycle time as an element. Although the two-part formula yields information that may be useful for advance analysis, most teams just starting out with OEE do not need that level of detail. For that reason, this book follows a simpler approach, used by Arno Koch of Blom Consultancy in his *OEE Toolkit* software, which compares actual output to the potential output if the machine were performing at its top speed.

PUBLISHER'S MESSAGE

This book is especially helpful when used with the *OEE Toolkit* software package (Productivity, 1999), which was developed by Arno Koch of Blom Consultancy to meet his clients' need for a simple and flexible approach to OEE tracking. The *OEE Toolkit* is an easy-to-use application for capturing OEE data and creating a wide range of reports from it. The manual that comes with the software teaches a people-centered approach to OEE measurement and reporting.

The overall equipment effectiveness measure is simple and universal. It is used to measure and improve equipment conditions in companies all over the world. We hope this book will tell you what you need to know to make your participation and use of OEE active and personally rewarding.

Acknowledgments

The development of *OEE for Operators* has been a team effort, and we greatly appreciate the contribution of everyone involved. The book was motivated by the approach to OEE developed by Arno Koch of Blom Consultancy and further supported by his *OEE Toolkit* software package. Content advisors included John Jacinto of Amtex and Bob Strout of Lemforder Co., as well as Productivity consultant John Monaco and *TPM Report* editor in chief Barry Shulak.

Lorraine Millard of Productivity managed the prepress production and manufacturing, with editorial assistance from Pauline Sullivan. Graphic illustrations were created by Guy Boster and Lee Smith, with cartoon illustrations by Guy Boster and Hannah Bonner. Cover composition was by Mark Weinstein of Productivity, with cover illustration by Gary Ragaglia of The Vision Group. Page composition was done by William H. Brunson Typography Services.

PUBLISHER'S MESSAGE

Finally, the Productivity staff wishes to acknowledge the good work of the many people who are in the process of implementing Total Productive Maintenance and using OEE in their own organizations. We welcome your feedback about this book, as well as input about how we can continue to serve your improvement efforts.

Steven Ott
President

Karen Jones
Productivity Development Team

Getting Started

The Purpose of This Book

Key Point

OEE for Operators was written to give you the information you need to participate in using the overall equipment effectiveness (OEE) measure in your workplace. You are a valued member of your company's team; your knowledge, support, and participation are essential to the success of any major effort in your organization.

The paragraph you have just read explains the author's purpose in writing this book. It also explains why your company may wish you to read this book. But why are *you* reading this book? This question is even more important. What you get out of this book largely depends on your purpose in reading it.

You may be reading this book because your team leader or manager asked you to do so. Or you may be reading it because you think it will provide information that will help you in your work. By the time you finish Chapter 1, you will have a better idea of how the information in this book can help you and your company measure equipment-related losses and plan how to improve equipment effectiveness.

What This Book Is Based On

BACKGROUND INFO

This book is about an approach for measuring equipment-related losses that limit the effectiveness of manufacturing equipment. Many of the methods discussed here were originally developed at companies working with the Japan Institute of Plant Maintenance, a pioneer in the approach known as Total Productive Maintenance, or TPM. Since 1988, Productivity, Inc. has made information about TPM approaches available in the United States through publications, events, training, and consulting. Today, top companies around the world are implementing TPM to sustain their competitive edge.

GETTING STARTED

Figure I-1. Two Ways to Use This Book

OEE for Operators draws on a wide variety of Productivity's book and training resources. Its aim is to present the main concepts and techniques of TPM and overall equipment effectiveness in a simple, illustrated format that is easy to read and understand. This book also complements the *OEE Toolkit* software package as a way to build a shared understanding among workteam members before they begin using OEE.

Two Ways to Use This Book

There are at least two ways to use this book:

1. As the reading material for a learning group or study group process within your company.
2. For learning on your own.

Your company may want to hold a series of learning group discussions based on this book. Managers may assign the book for background reading when the company uses the *OEE Toolkit* software package. Or, you may read this book for individual learning without formal group discussion.

GETTING STARTED

How to Get the Most Out of Your Reading
Becoming Familiar with This Book as a Whole

There are a few steps you can follow to make it easier to absorb the information in this book. Take as much time as you need to become familiar with the material. First, get a "big picture" view of the book by doing the following:

How-to Steps

1. Scan the Contents (pages v through vii) to see how *OEE for Operators* is arranged.

2. Read the rest of this section for an overview of the book's contents.

3. Flip through the book to get a feel for its style, flow, and design. Notice how the chapters are structured and glance at the pictures.

Becoming Familiar with Each Chapter

After you have a sense of the structure of *OEE for Operators*, prepare yourself to study one chapter at a time. For each chapter, we suggest you follow these steps to get the most out of your reading:

How-to Steps

1. Read the "Chapter Overview" on the first page to see where the chapter is going.

2. Flip through the chapter, looking at the way it is laid out. Notice the bold headings and the key points flagged in the margins.

3. Now read the chapter. How long this takes depends on what you already know about the content, and what you are trying to get out of your reading. Enhance your reading by doing the following:

   • Use the margin assists to help you follow the flow of information.

   • If the book is your own, use a highlighter to mark key information and answers to your questions about the material. If the book is not your own, take notes on a separate piece of paper.

   • Answer the "Take Five" questions in the text. These will help you absorb the information by reflecting on how you might apply it at work.

4. Read the "Chapter Summary" to confirm what you have learned. If you don't remember something in the summary, find that section in the chapter and review it.

5. Finally, read the "Reflections" questions at the end of the chapter. Think about these questions and write down your answers. Find an experienced person to ask if you find a topic confusing.

GETTING STARTED

Figure I-2. Giving Your Brain a Framework for Learning

How a Reading Strategy Works

When reading a book, many people think they should start with the first word and read straight through until the end. This is not usually the best way to learn from a book. The steps described on page xv are a strategy for making your reading easier, more fun, and more effective.

Key Point

Reading strategy is based on two simple points about the way people learn. The first point is this: *It is difficult for your brain to absorb new information if it does not have a structure to place it in.* As an analogy, imagine trying to build a house without first putting up a framework.

Like building a frame for a house, you can give your brain a framework for the new information in the book by getting an overview of the contents and then flipping through the materials. Within each chapter, you repeat this process on a smaller scale by reading the overview, key points, and headings before reading the text.

Key Point

The second point about learning is this: *It is a lot easier to learn if you take in the information one layer at a time, instead of trying to absorb it all at once.* It's like finishing the walls of a house: First you lay down a coat of primer. When it's dry, you apply a coat of paint, and later a final finish coat.

GETTING STARTED

Using the Margin Assists

As you've noticed by now, this book uses small images called *margin assists* to help you follow the information in each chapter. There are six types of margin assists:

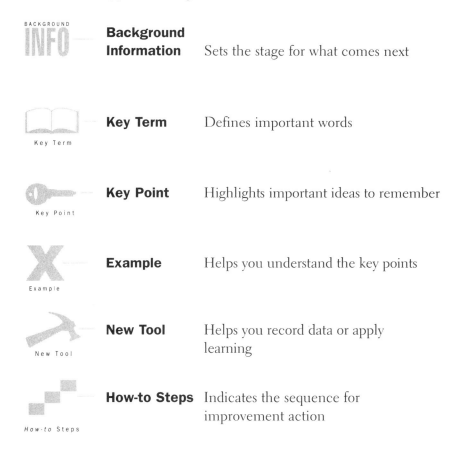

Background Information — Sets the stage for what comes next

Key Term — Defines important words

Key Point — Highlights important ideas to remember

Example — Helps you understand the key points

New Tool — Helps you record data or apply learning

How-to Steps — Indicates the sequence for improvement action

Overview of the Contents

Getting Started (pages xv–xx)

This is the section you're reading now. It explains the purpose of *OEE for Operators* and how it was written. Then it shares tips for getting the most out of your reading. Finally, it presents this overview of each chapter.

GETTING STARTED

Chapter 1. About TPM and OEE (pages 1–8)

Chapter 1 introduces and defines Total Productive Maintenance and overall equipment effectiveness. It explains reasons why OEE is an important measure to track and describes the role of the shopfloor team in collecting and using OEE data.

Chapter 2. Understanding Equipment-Related Losses (pages 9–23)

Chapter 2 describes the three elements of OEE and links them to the main types of losses that lower equipment effectiveness.

Chapter 3. Measuring OEE (pages 25–38)

Chapter 3 offers guidance in measuring overall equipment effectiveness, including collecting and processing data and using the resulting information on the shop floor. It tells about defining what data to measure for the OEE calculation, doing the calculation, and storing the data so you can report the information in different ways.

Chapter 4. Improving OEE (pages 39–56)

Chapter 4 covers essential approaches for improving overall equipment effectiveness. Topics include 5 Why analysis, the autonomous maintenance and focused improvement pillars of TPM, specific approaches for setup and defect losses, and the advanced P-M analysis approach for chronic problems.

Chapter 5. Reflections and Conclusions (pages 57–63)

Chapter 5 presents reflections on and conclusions to this book. It also describes opportunities and specific resources for further learning about OEE, TPM, and related techniques.

Chapter 1
About TPM and OEE

CHAPTER OVERVIEW

What Is TPM?

What Is OEE and Why Is It Important?
Quantity Over Time Is Only Part of OEE

Effectiveness Focuses on the Equipment, Not the Person

The Purpose of Measurement Is Improvement

The Role of the Shopfloor Team in Using OEE

In Conclusion
Summary

Reflections

CHAPTER 1

What Is TPM?

Overall equipment effectiveness (OEE) is a key measurement in the improvement approach called Total Productive Maintenance (TPM). Before you begin learning about OEE, it is useful to understand a little bit about TPM.

TPM is a companywide approach for improving the effectiveness and longevity of machines. It is key to lean manufacturing because it attacks major wastes in production operations. Developed originally to help a supplier meet the stringent requirements of the Toyota Production System, TPM is used today in companies around the world to improve the capability of their equipment.

TPM has a number of waste-reduction goals, including equipment restoration and maintenance of standard operating conditions. TPM methods also improve equipment systems, operating procedures, and maintenance and design processes to avoid future problems.

The main strategies used in TPM are often referred to as "pillars" that support the smooth operation of the plant. Figure 1-1 summarizes the activities in eight basic pillars of TPM.

The overall equipment effectiveness measure is important to many of the TPM pillars, but is probably most important to the first four pillars in the figure. This is because these pillars can directly influence OEE through daily operations, maintenance, or improvement activities.

Pillar	Activities
Focused equipment and process improvement	Measurement of equipment- or process-related losses and specific improvement activities to reduce the losses.
Autonomous maintenance	Operator involvement in regular cleaning, inspection, lubrication, and learning about equipment to maintain basic conditions and spot early signs of trouble.
Planned maintenance	A combination of preventive, predictive, and proactive maintenance to avoid losses, and planned responses to fix breakdowns quickly.
Quality maintenance	Activities to manage product quality by maintaining optimal operating conditions.
Early equipment management	Methods to shorten the lead time for getting new equipment online and making defect-free products.
Safety	Safety training; integration of safety checks, visual controls, and mistake-proofing devices in daily work.
Equipment investment and maintenance prevention design	Purchase and design decisions informed by costs of operation and maintenance during the machine's entire life cycle.
Training and skill building	A planned program for developing employee skills and knowledge to support TPM implementation.

Figure 1-1. Basic Pillars of TPM

CHAPTER 1

What Is OEE and Why Is It Important?

Manufacturing companies are in business to make money, and they make money by adding value to materials to make products the customers want.

Key Term

Most companies use machines to add value to the products. To add value effectively, it is important to run the machines effectively, with as little waste as possible. *Overall equipment effectiveness is a measurement used in TPM to indicate how effectively machines are running.*

What do we mean by overall equipment *effectiveness*? Many people are familiar with the idea of "efficiency," which usually reflects the quantity of parts a machine or a person can produce in a certain time. OEE is different from efficiency in several ways.

Quantity Over Time Is Only Part of OEE

A machine's overall effectiveness includes more than the quantity of parts it can produce in a shift. When we measure overall equipment effectiveness, we account for efficiency as one factor:

Key Term

- *Performance*: a comparison of the actual output with what the machine should be producing in the same time.

In addition to performance, however, OEE includes two other factors:

Key Terms

- *Availability*: a comparison of the potential operating time and the time in which the machine is actually making products.
- *Quality*: a comparison of the number of products made and the number of products that meet the customer's specifications.

Key Point

When you multiply performance, availability, and quality, you get the overall equipment effectiveness, which is expressed as a percentage. OEE gives a complete picture of the machine's "health"—not just how fast it can make parts, but how much the potential output was limited due to lost availability or poor quality (see Figure 1-2). In Chapter 2 we will look more closely at these three elements and how they work together.

ABOUT TPM AND OEE

Figure 1-2. Efficiency Is Not the Same as Effectiveness

Effectiveness Focuses on the Equipment, Not the Person

Unlike some uses of the efficiency measure, OEE monitors the machine or process that adds the value, not the operator's productivity. *When we measure OEE, we look at how well the* equipment or process *is working*.

Key Point

The Purpose of Measurement Is Improvement

Measuring OEE is not an approach for criticizing people. It is strictly about improving the equipment or process. *Used as an impartial daily snapshot of equipment conditions, OEE promotes openness in information sharing and a no-blame approach in handling equipment-related issues.*

Key Point

These key differences highlight the importance of OEE as a balanced measure that helps support improvement and profitability.

TAKE FIVE

Take five minutes to think about these questions and to write down your answers:

- Does your company currently measure each machine's efficiency? Its available running time? Its quality rate?

CHAPTER 1

Figure 1-3. Collecting Data and Turning It into Information

The Role of the Shopfloor Team in Using OEE

Key Point

This book is written for you, the shopfloor employee, because you have a big stake in the health of the production equipment. As operators, you manage the equipment that adds value to the product. When the machines break down, run too slowly, or produce defects, you have to work longer and harder to make up for the problems. The pressure these problems creates is a good incentive to measure them and start improving them.

Key Point

What's more, your daily work with the machines puts you in the best position to monitor their problems. You know how long a machine is shut down for setup, or when minor stoppages get in the way of high-speed operation, or when you have to run slower to avoid defects. In many cases, you already track the data that will be used to calculate the overall equipment effectiveness.

Key Point

Sharing information on the plant floor through graphs and discussion is the heart of TPM (see Figure 1-3). The OEE information isn't useful when it is locked away in an office. The best approach for applying OEE gives operators a leading role in gathering daily data, converts the data into useful information, and applies the information in the workplace to support improvement.

In Conclusion

SUMMARY

- Overall equipment effectiveness (OEE) is a key measurement in the improvement approach called Total Productive Maintenance (TPM).
- TPM is a companywide approach for improving the effectiveness and longevity of machines.
- TPM has a number of waste-reduction goals, including equipment restoration and maintenance of standard operating conditions. TPM methods also improve equipment systems, operating procedures, and maintenance and design processes to avoid future problems.
- Overall equipment *effectiveness* is a measurement used in TPM to indicate how effectively machines are running.
- Overall equipment effectiveness is not the same as efficiency, which usually means how many parts a machine or a person can produce in a certain time. OEE is different in several ways.
  - *Quantity over time is only one part of OEE.*
    A machine's overall effectiveness includes more than the quantity of parts it can produce in a shift. OEE includes efficiency as one factor—performance—but also two other factors—availability and quality. When you multiply performance, availability, and quality, you get the overall equipment effectiveness, which is expressed as a percentage.
  - *Effectiveness focuses on the equipment or process, not the person.*
    When we measure OEE, we pay attention to how well the equipment or process is performing, not the operator's productivity.
  - *The purpose of measurement is improvement.*
    Used as an impartial daily snapshot of the equipment, OEE promotes openness in information sharing and a no-blame approach in handling equipment-related issues.

- This book is written for you, the shopfloor employee, because you have a big stake in the health of the production equipment. What's more, your daily work with the machines puts you in the best position to monitor their problems.

- Sharing information on the plant floor through graphs and discussion is the heart of TPM.

REFLECTIONS

Now that you have completed this chapter, take five minutes to think about these questions and to write down your answers:

- What did you learn from reading this chapter that stands out as particularly useful or interesting?

- Do you have any questions about the topics presented in this chapter? If so, what are they?

Chapter 2

Understanding Equipment-Related Losses

CHAPTER OVERVIEW

Losses Reduce Overall Equipment Effectiveness

Visualizing OEE and the Losses
    Availability
    Performance
    Quality

Availability: Downtime Losses
    Failures and Repairs
    Setup Time
    Other Losses to Availability

Performance: Speed Losses
    Reduced Operating Speed
    Minor Stoppages

Quality: Defect Losses
    Scrap and Rework
    Startup Loss

In Conclusion
    Summary
    Reflections

CHAPTER 2

Figure 2-1. Ideal and Actual Effectiveness

Losses Reduce Overall Equipment Effectiveness

What makes machines less effective than they could be? The ideal, totally effective machine could run all the time (or whenever needed). It could maintain its maximum or standard speed all the time. It would never make defective products.

But most machines aren't ideal. They cannot run continuously. They cannot maintain maximum speed without problems. And they make defects.

Key Term

These problems are familiar forms of waste—they don't add value to the products. They reduce a machine's effectiveness, as measured by the OEE. *The conditions that cause these machine problems are called equipment-related losses.* Understanding the different types of equipment-related losses will give you a framework for applying OEE and participating in improvement activities to reduce the losses.

UNDERSTANDING EQUIPMENT-RELATED LOSSES

Key Term

The equipment-related losses that are important for OEE are linked to the three basic elements measured in OEE: availability, performance, and quality. *Traditional TPM approaches track "Six Major Losses":*

Availability: Downtime losses	Performance: Speed losses	Quality: Defect losses
• Failures	• Minor stoppages	• Scrap and rework
• Setup time	• Reduced operating speed	• Startup loss

Although some companies link individual losses to different OEE categories, or add other losses that are especially significant for their operations, this basic framework is a useful starting point for many companies. Figure 2-2 on the next page gives a visual image of the way in which these losses reduce the overall equipment effectiveness of a machine.

TAKE FIVE

Take five minutes to think about this question and to write down your answer:

- What are some of the situations that keep your machines from running at an ideal level of effectiveness?

CHAPTER 2

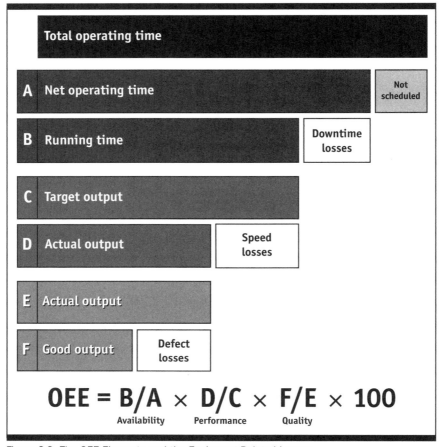

Figure 2-2. The OEE Elements and the Equipment-Related Losses

Visualizing OEE and the Losses

Key Point

Figure 2-2 makes it easy to see how *OEE is derived from the three elements, expressed as fractions.* Each pair of bars stands for one of the fractions—availability (B/A), performance (D/C), and quality (F/E). The fractions are often multiplied by 100 to turn them into percentages or rates.

Availability

Bars A and B represent availability. Unscheduled time shortens the total operating time,* leaving net operating time (A). But the

---

*Companies count this time in different ways, but for this discussion, we subtract these periods from the total operating time.

UNDERSTANDING EQUIPMENT-RELATED LOSSES

machine is frequently down during some of that time, usually due to breakdowns and setup. Subtracting that downtime leaves the running time (B) in which the machine is making product.

Example: $\dfrac{\text{running time}}{\text{net operating time}} = \dfrac{300 \text{ minutes}}{400 \text{ minutes}} = .75 \text{ availability} (\times 100 = 75\%)$

Performance

Bars C and D represent performance. During the running time, the machine could produce a target output quantity (C) if it ran at its designed speed the whole time. But losses such as minor stoppages and reduced operating speed lower the actual output (D).

Example: $\dfrac{\text{actual output}}{\text{target output}} = \dfrac{12{,}000 \text{ parts}}{20{,}000 \text{ parts}} = .60 \text{ performance} (\times 100 = 60\%)$

Quality

Bars E and F represent quality. Of the actual output (E), most of the product is good output (F). But usually some output falls short of the specified quality and must be scrapped or reworked. Scrap is often produced during machine startup as well, lowering the yield from the materials.

Example: $\dfrac{\text{good output}}{\text{actual output}} = \dfrac{11{,}760 \text{ parts}}{12{,}000 \text{ parts}} = .98 \text{ quality} (\times 100 = 98\%)$

Figure 2-2 shows how *losses to availability, performance, and quality compound to reduce the amount of good output a machine can produce during a shift*. You can improve quality to raise the quantity of good output a little bit—but the total quantity won't rise dramatically unless you also improve both performance and availability.

The formula at the bottom of Figure 2-2 shows how to multiply the three elements to get the OEE.

Example: .75   ×   .60   ×   .98   × 100 = 44% OEE
     (availability) (performance) (quality)

The rest of this chapter will look more closely at the losses associated with these elements.

13

CHAPTER 2

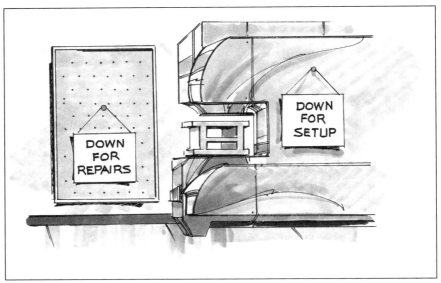

Figure 2-3. Downtime Losses—Failures and Setup

Availability: Downtime Losses

Failures

Key Point

Availability is reduced by equipment failures, which are a common occurrence in many plants. Machines used for production generally have lots of moving parts and various subsystems in which things can go wrong. When they do, the machine breaks down—and stays down until repairs are completed.

Key Point

Many of the causes of machine failure give warning signs before the machine actually breaks. In Chapter 4 we will look at how autonomous maintenance activities can help spot early trouble signs in time to prevent major breakdowns.

Setup Time

Key Point

Availability is also reduced by the time it takes to set up the machine for a different product. In addition to changing the value-adding parts, a changeover requires some preparation or make-ready. It may involve cleaning and making adjustments to the machine to get stable quality in the next product. Too often, it also involves running around to find tools, parts, or people. We will consider an approach for reducing setup time in Chapter 4.

14

UNDERSTANDING EQUIPMENT-RELATED LOSSES

Figure 2-4. Downtime Losses—Cutting Tool Loss and Startup Loss

Other Losses to Availability

Failures and setup losses were the original losses counted as downtime that reduces availability. Some companies also track other losses as downtime, depending on what losses they are trying to improve. Cutting tool loss, startup loss, and time not scheduled for production are three other losses tracked as downtime at some plants.

Cutting Tool Loss

Breakage of cutting tools during production causes unplanned downtime while the tool is replaced. Although this is technically a subset of failure and repair losses, some companies track it separately because of the potential for injury and product defects, as well as the cost of tool replacement. Planned maintenance and autonomous maintenance activities help reduce these losses.

Startup Loss

Startup loss is traditionally included as a defect loss, since its essence is the production of defective products during startup. However, *startup loss involves lost time until good production can be stabilized,* so it is logical to subtract it from available time as well.

15

CHAPTER 2

Figure 2-5. Downtime Losses—Unscheduled Time

Time Not Scheduled for Production

In some companies, when machines are stopped for meetings, preventive maintenance, or breaks, the time is considered "not scheduled" and is not counted in the availability rate (see Figure 2-5). Other companies recognize that even necessary activities like these reduce the available production time. They may decide to consider time "not scheduled" as a downtime loss that lowers the availability rate.

Key Point

Counting unscheduled time as a loss can encourage creative ideas for reducing the loss—without eliminating the activity. For example, after measuring the production time lost from scheduled breaks, employees at one company developed a plan to alternate their breaks and briefly cover each other's stations.

UNDERSTANDING EQUIPMENT-RELATED LOSSES

Likewise, some companies count offline time for preventive maintenance as downtime. Again, the point is to reduce the time loss, not to eliminate the activity.

TAKE FIVE

Take five minutes to think about these questions and to write down your answers:

- How much time is lost each month due to failures and repairs in your area?
- How much time is spent each month on setup and make-ready in your area?
- Would you count other time losses for OEE purposes? Why or why not?

CHAPTER 2

Figure 2-6. Speed Losses: Minor Stoppages

Performance: Speed Losses

Reduced Operating Speed

Key Point

Machines often run at speeds slower than they were designed to run. One reason for slower operation is unstable product quality at the designed speed. In other cases, people don't realize that the equipment is designed to run faster. We will look in Chapter 3 at how to determine speed for the OEE calculation.

Minor Stoppages

Key Term

Minor stoppages are events that interrupt the production flow without actually making the machine fail. They often occur on automated lines, for example when product components snag on the conveyor (see Figure 2-6).

Key Point

Minor stoppages can make it impossible to run automated equipment without someone to monitor it. *These stoppages may seem like petty annoyances, but they add up to big losses at many plants.*

UNDERSTANDING EQUIPMENT-RELATED LOSSES

Minor stoppages last only a few seconds, so we don't try to log the time lost. Instead, we include them in performance losses that reduce the product output. We will look at approaches for reducing speed losses in Chapter 4.

TAKE FIVE

Take five minutes to think about these questions and to write down your answers:

- Do you know the designed speed of the machines in your area?
- Do minor stoppages happen in your area? What causes them?

CHAPTER 2

Figure 2-7. Defect Losses: Scrap, Rework, and Startup Loss

Quality: Defect Losses

Scrap and Rework

Products that do not meet customer specifications are a familiar loss. Clearly, scrap that cannot be reused is a waste of materials. Even when products can be reworked, the effort spent to process them twice is a waste.

Startup Loss

Key Point

Many machines take time to reach the right operating conditions at startup. In the meantime, they may turn out defective products while operators test for stable output. Some companies simply include this startup loss in scrap and rework; others single it out as a specific loss to track.*

---

*As mentioned in the section on downtime losses, some companies also single out the startup period before the first good product as a special type of downtime to track.

UNDERSTANDING EQUIPMENT-RELATED LOSSES

Quality problems happen when the optimum conditions do not exist at the moment when a person or machine works on the product. In Chapter 4 we will look at a method for preventing defects by checking and controlling the necessary conditions.

> **TAKE FIVE**
>
> Take five minutes to think about these questions and to write down your answers:
> - What is the defect rate for machines in your area? Do you think this can be reduced?
> - Are startup losses a significant problem in your area?

In Conclusion

SUMMARY

- The ideal, totally effective machine would run all the time (or whenever needed), at maximum or standard speed, with no quality problems. But most machines can't meet these ideal conditions. They can't run continuously or at maximum speed; they experience minor stoppages, and they make defective parts.

- These problems reduce a machine's effectiveness, as measured by the OEE. The conditions that cause these problems are called equipment-related losses. Linked to the three basic elements of OEE, they include the traditional "Six Major Losses":

Availability:	**Performance:**	**Quality:**
Downtime losses	**Speed losses**	**Defect losses**
• Failures	• Minor stoppages	• Scrap and rework
• Setup time	• Reduced operating speed	• Startup loss

- Although some companies link individual losses to different OEE categories, or add other losses that are especially significant for their operations, this basic framework is a useful starting point.

- OEE is derived from the three elements, expressed as fractions. The fractions are often multiplied by 100 to turn them into percentages or rates. Losses to these three elements reduce the amount of good output a machine can produce during a shift.

- Downtime losses affect availability. Failures and setup time are common losses tracked.
  - Some companies also track other losses as downtime, depending on what they are trying to improve. Cutting tool loss, startup loss, and time not scheduled for production are three other losses sometimes tracked as downtime.

- Speed losses affect performance. Minor stoppages and operation at reduced speed are often measured as speed losses.

UNDERSTANDING EQUIPMENT-RELATED LOSSES

- Defect losses affect quality. They include scrap and rework when products do not meet customer specifications.
  - Also, many machines turn out defective products during startup while operators test for stable output. Some companies include this loss in scrap and rework; others single it out as a specific loss to track.

REFLECTIONS

Now that you have completed this chapter, take five minutes to think about these questions and to write down your answers:

- What did you learn from reading this chapter that stands out as particularly useful or interesting?

- Do you have any questions about the topics presented in this chapter? If so, what are they?

Chapter 3
Measuring OEE

CHAPTER OVERVIEW

Closing the Feedback Loop

Collecting OEE Data
    Defining What to Measure
    Making Data Collection Simple

Processing OEE Data
    The OEE Calculation
    Storing OEE Data

Reporting OEE Results

In Conclusion
    Summary
    Reflections

CHAPTER 3

Key Point

Measuring overall equipment effectiveness is an important way to monitor which losses are reducing the effectiveness of your machines. *By tracking OEE on a regular basis, you can spot patterns and influences that cause problems for production equipment. Furthermore, measuring OEE allows you to see the results of your efforts to help the machines run better.* This chapter offers guidance in measuring overall equipment effectiveness, including collecting and processing OEE data and reporting OEE results.

Closing the Feedback Loop

The process of measuring and applying OEE data should involve the people who use the machines. As operators, you are more familiar than other people with the equipment you operate, and you have a stake in helping it run well. Therefore it's logical for you to take part in collecting the data for calculating OEE.

Key Point

Just as important as being involved in data collection is receiving feedback on OEE results. An OEE chart cannot promote improvement if it doesn't get back to the shop floor. OEE is living information for improving equipment effectiveness. It should not be buried away in an office.

Collecting OEE Data

Defining What to Measure

Key Point

Before you can begin applying OEE, you need to decide what machine and product data you will measure for the calculation. The basic items you will measure are the losses that reduce availability, performance, and quality. These will vary from plant to plant, but the Six Major Losses described in Chapter 2 give a good framework to start from.

Downtime Losses

Key Point

Downtime losses (lost availability) are measured in units of time (Figure 3-1). They include

- failure and repair time
- setup and adjustment time
- other time losses that reduce availability

MEASURING OEE

Figure 3-1. Failure and Setup Losses Are Measured as Time Losses

Failure and repair time includes all of the downtime until the machine makes the next good product. Some plants lump all breakdowns into one category; other plants may create several categories to distinguish between different types or causes of machine failures. The main thing is to standardize your approach so everyone can measure a failure event the same way.

Setup and adjustment time includes the time between the last good piece of product A and the first good piece of product B.

Other time losses include startup losses—similar to setup time losses—and any nonscheduled time the team chooses to subtract from the available time.

TAKE FIVE

Take five minutes to think about these questions and to write down your answers:

- What types of information about your machine's operation do you currently track?
- What types of downtime losses do you think your work area would track for OEE?

CHAPTER 3

Figure 3-2. Minor Stoppages and Reduced Operating Speed Are Measured as Output Reductions

Speed Losses

Speed losses (lost performance) are measured in units of product output (see Figure 3-2). You probably already track your output quantity. For OEE, you look at the difference between the actual output and the potential output if the machine consistently ran at the designed speed, or at the standard optimum speed for each product.

Speed losses include minor stoppages as well as reduced operating speed. Although minor stoppages are "events" like mini-breakdowns, they often occur so frequently that it is not practical to record the time lost during many frequent stoppages. For that reason, many companies monitor minor stoppages by tracking the output reduction they cause.

To compare the actual output rate (machine speed) with the output rate at the designed speed, you have to know what the designed speed is. If this speed does not appear in the machine's documentation, you will need to set a standard, such as the fastest known speed at which the machine can run (this may vary for different products).

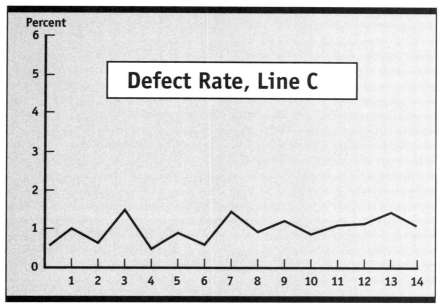

Figure 3-3. Scrap and Startup Losses Are Measured as Defective Output Compared to Total Output

Defect Losses

Key Point

Defect losses (lost quality) are also measured in units of product output. This time, you are looking at the difference between the total actual output and the output that meets customer specifications (see Figure 3-3).

Defect losses include products that can be reworked as well as outright scrap. First-pass quality is the goal.

Making Data Collection Simple

The purpose of tracking OEE is not to make extra paperwork for operators. Most likely you are already collecting a lot of the data required for the OEE calculation. *One well-designed form can make it easy to log the OEE data as well as other data you need to register during daily production.*

CHAPTER 3

Example

Figure 3-4 (pages 31–32) shows a sample data collection form. Its creators used a simple approach for logging time losses by shading the boxes on Side A to indicate where downtime occurred. Performance and quality data go on Side B.

> **TAKE FIVE**
>
> Take five minutes to think about these questions and to write down your answers:
> - Which type of loss most affects your production equipment?
> - How would you change your current data collection forms to include OEE data?

MEASURING OEE

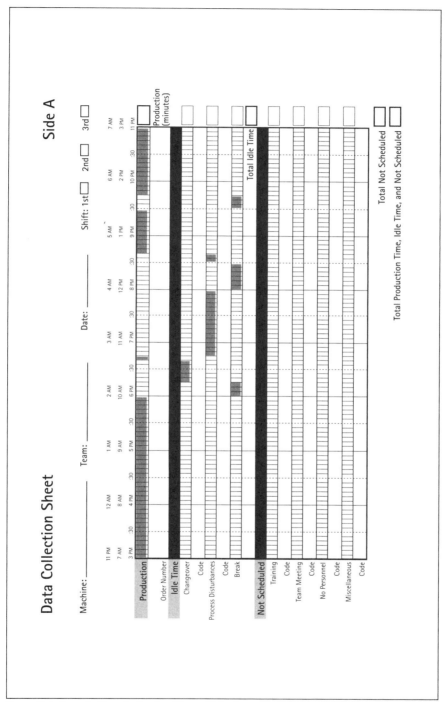

Figure 3-4. Sample Data Collection Sheet (Side A)
Source: Arno Koch, Blom Consultancy, *User's Guide* for *OEE Toolkit* software (Productivity, 1999).

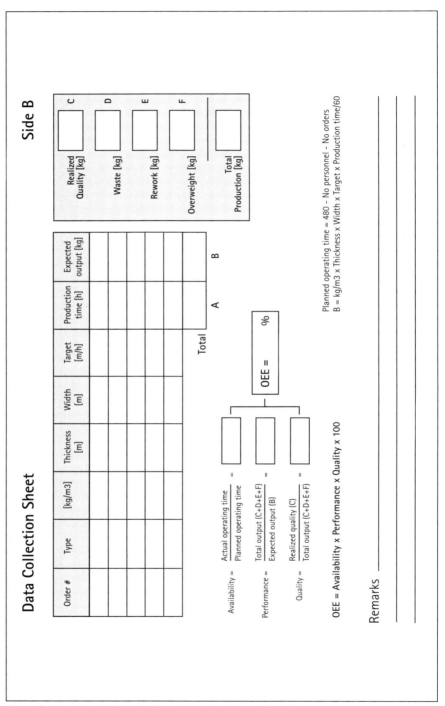

Figure 3-4 (continued). Sample Data Collection Sheet (Side B)

Source: Arno Koch, Blom Consultancy, *User's Guide* for *OEE Toolkit* software (Productivity, 1999).

MEASURING OEE

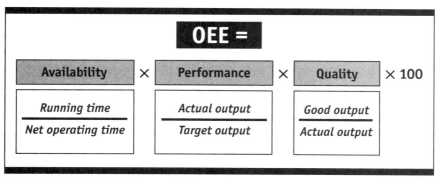

Figure 3-5. The OEE Calculation and Its Elements

Processing OEE Data

After you collect data for OEE, you need to process the data to turn it into useful information. This involves doing the calculation, and also storing your data in a way that allows you to draw different types of information from it.

The OEE Calculation

OEE is calculated by multiplying availability, performance, and quality (multiplied by 100 to give a percentage rate).

$$OEE\ rate = Availability \times Performance \times Quality \times 100$$

Let's review the equations for the individual elements of OEE.

$$Availability = \frac{Running\ time}{Net\ operating\ time}$$

The running time is the net operating time minus the downtime losses you decide to measure.

$$Performance = \frac{Actual\ output}{Target\ output}$$

For the OEE calculation, the target output is the quantity the machine would produce if it operated at its designed speed during the running time (see Figure 3-5).

$$Quality = \frac{Good\ output}{Actual\ output}$$

CHAPTER 3

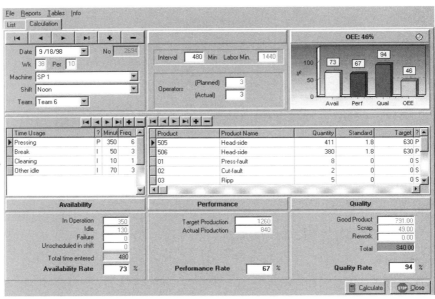

Figure 3-6. Software Can Be Helpful for OEE Calculation and Data Storage

Source: Sample data entered in *OEE Toolkit* software application (Arno Koch, Blom Consultancy; Productivity, 1999).

Storing OEE Data

OEE is most valuable when you collect data and do the calculation on a regular basis. Tracking OEE at set intervals over time allows you to see patterns that give clues for improvement.

Key Point

It is important to have a system in place to store your OEE data. Manual charting of the basic rates is a good place to start, but it limits the information you can pull out of the data. Software can be a helpful tool for automating the calculation and storing the data for use in several types of graphs (see Figure 3-6).

TAKE FIVE

Take five minutes to think about this question and to write down your answer:

- What kind of data storage system would you want to use for your OEE measurements?

MEASURING OEE

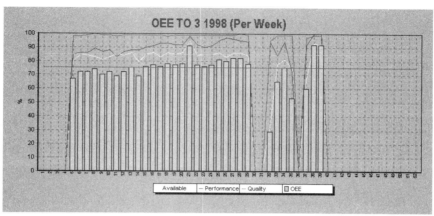

Figure 3-7. OEE Trend Chart

Source: Sample data entered in *OEE Toolkit* software application (Arno Koch, Blom Consultancy; Productivity, 1999).

Reporting OEE Results

Sharing OEE information is critical for reducing equipment-related losses. Operators—the people who are closest to the equipment—need to be aware of OEE results. Reporting OEE information on charts in the workplace is a key to improving future results.

The graph in Figure 3-6 shows a typical machine's OEE chart for one shift. By tracking this data over time, you can see the OEE trends for the machine, as shown in Figure 3-7.

Key Point

The few pieces of data you collect to track OEE can give a lot of other information about the machine, answering such questions as

- Are we improving over time?
- What are the biggest downtime problems?
- When did an incident occur?
- How was quality over the last month?
- How are we utilizing the equipment?
- What is our mean time between failures, failure rate and frequency, and mean time to repair?

Figures 3-8 and 3-9 show sample reports from OEE data.

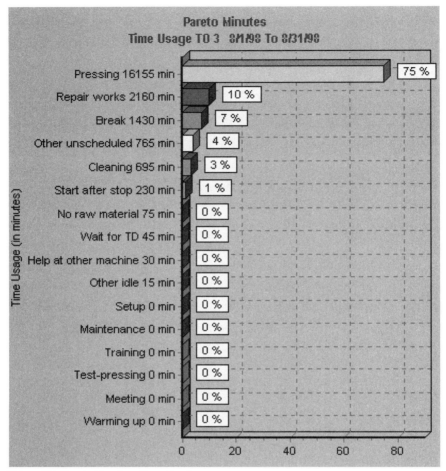

Figure 3-8. Pareto Chart on Time Usage
Source: Sample data entered in *OEE Toolkit* software application (Arno Koch, Blom Consultancy; Productivity, 1999).

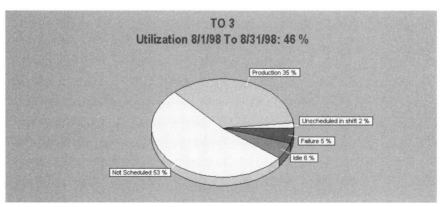

Figure 3-9. Utilization Chart
Source: Sample data entered in *OEE Toolkit* software application (Arno Koch, Blom Consultancy; Productivity, 1999).

In Conclusion

SUMMARY

- Tracking OEE helps you spot patterns and influences of equipment problems and allows you to see the results of your improvement efforts.

- The process of measuring and applying OEE data should involve the people who use the machines. Operators should also receive feedback on OEE results.

- Before you begin applying OEE, you need to decide what to measure for the calculation. The basic items you will measure are the losses that reduce availability, performance, and quality. These will vary from plant to plant, but the Six Major Losses provide a good starting framework.

- Downtime losses (lost availability) are measured in units of time. They include
    - failure and repair time
    - setup and adjustment time
    - other time losses that reduce availability

- Speed losses (lost performance) are measured in units of product output. You will look at the difference between the actual output and the potential output if the machine consistently ran at the designed speed or the standard optimum speed for each product.

- Defect losses (lost quality) are also measured in units of product output. Here you are looking at the difference between the total actual output and the output that meets customer specifications.

- The purpose of tracking OEE is not to make extra paperwork. A well-designed form can make it easy to log the OEE data as well as other data you need to register during daily production.

- After you collect data for OEE, you need to process the data to turn it into useful information. This involves doing the calculation, and also storing your data in a way that allows you to draw different types of information from it.

- OEE is calculated by multiplying availability, performance, and quality (multiplied by 100 to get a percentage rate).

  $OEE\ rate = Availability \times Performance \times Quality \times 100$

  $Availability = \dfrac{Running\ time}{Net\ operating\ time}$

  $Performance = \dfrac{Actual\ output}{Target\ output}$

  $Quality = \dfrac{Good\ output}{Actual\ output}$

- Tracking OEE at set intervals over time allows you to see patterns that give clues for improvement.

- It is important to have a system in place to store your OEE data. Software can be helpful for automating the calculation and storing the data for use in reports.

- Reporting the results on charts in the workplace is a key to improving future results. The few pieces of data you collect to track OEE can give a lot of other information about the machine.

REFLECTIONS

Now that you have completed this chapter, take five minutes to think about these questions and to write down your answers:

- What did you learn from reading this chapter that stands out as particularly useful or interesting?

- Do you have any questions about the topics presented in this chapter? If so, what are they?

Chapter 4
Improving OEE

CHAPTER OVERVIEW

5 Why Analysis

Autonomous Maintenance

Focused Equipment and Process Improvement

Quick Changeover
- Stage 1: Separate Internal and External Setup
- Stage 2: Convert Internal Setup to External Setup
- Stage 3: Streamline All Aspects of Setup

ZQC (Mistake-Proofing)
- Poka-Yoke Systems

P-M Analysis

In Conclusion
- Summary
- Reflections

CHAPTER 4

Figure 4-1. OEE Tells the Current State of the Equipment

We measure OEE to monitor the condition of the equipment—similar to what a nurse learns about your condition when he or she takes a temperature or listens to a heartbeat (see Figure 4-1). By comparing yesterday's or last week's result, we can see whether the condition has improved or become worse. As an operator, you play an important role in TPM because you are in the best position to monitor machine conditions during operation.

Key Point

The point of using the OEE measure is to drive improvement. When you first begin tracking OEE, the rate may be very low. This is not totally bad, because it means there is a big opportunity to improve. It is much easier to improve a low OEE rate than a high one, since people tend to eliminate the obvious wastes and problems at the beginning.

Standardization is the first step in improvement. OEE is a tool for standardizing the way you measure effectiveness. This standardized approach provides a baseline that helps you see where to focus improvement efforts.

Key Point

Some improvement may happen just from the awareness that develops when you start measuring OEE. *Sustained improvement, however, requires a dedicated approach, with management support.* This chapter explores several approaches that can help improve OEE.

IMPROVING OEE

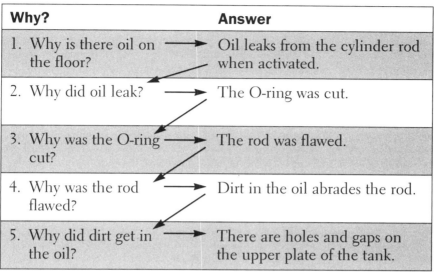

Figure 4-2. 5 Why Analysis

5 Why Analysis

Have you ever had the experience in which someone fixed a machine problem, but the same problem happened again after a short time? In such cases, it often turns out that people have been treating the symptoms of the problem, but not dealing with its real, root cause. Until we address the root cause, the same problem will keep returning.

New Tool

5 Why analysis is a useful tool that brings us closer to the root cause. As its name suggests, *5 Why analysis involves repeatedly asking "why?" about the problem* (it could be more or less than five times, depending on the situation). This leads us to look beyond the immediate effect—such as a broken drive belt—to see the factors that might be causing the effect—such as flaws on the pulley that make the belt wear out too soon.

Figure 4-2 shows an example of using 5 Why analysis.

> ### TAKE FIVE
>
> Take five minutes to think about these questions and to write down your answers:
>
> - Is there a typical situation in your workplace where people "fix the symptoms"? What do you think is the root cause, and what would you do about it?

CHAPTER 4

Figure 4-3. Autonomous Maintenance Involves Everyone!

Autonomous Maintenance

Key Term

Autonomous maintenance refers to activities carried out by shopfloor teams in cooperation with maintenance staff to help stabilize basic equipment conditions and spot problems early. Autonomous maintenance is one of the pillars of TPM. It changes the old view that operators just run machines and maintenance people just fix them. Operators have valuable knowledge and skill that can help keep equipment from breaking down.

In autonomous maintenance, operators learn how to clean the equipment they use every day, and how to inspect for trouble signs as they clean (see Figure 4-3). They may also learn basic lubrication routines, or at least how to check for adequate lubrication. They learn simple methods to reduce contamination and keep the equipment cleaner. Ultimately, they learn more about the various operating systems of the equipment and may assist technicians with repairs.

Key Point

Autonomous maintenance activities are like exercise and regular health checkups for machines. Along with preventive maintenance, they help raise OEE by maintaining proper operating conditions, and stabilize it by detecting abnormalities before they turn into losses.

> **Step 1.** Conduct initial cleaning and inspection.
>
> **Step 2.** Eliminate sources of contamination and inaccessible areas.
>
> **Step 3.** Develop and test provisional cleaning, inspection, and lubrication standards.
>
> **Step 4.** Conduct general inspection training and develop inspection procedures.
>
> **Step 5.** Conduct general inspections autonomously.
>
> **Step 6.** Apply standardization and visual management throughout the workplace.
>
> **Step 7.** Conduct ongoing autonomous maintenance and advanced improvement activities.

Figure 4-4. Autonomous Maintenance Activities

Key Point

Autonomous maintenance is, at its heart, a team-based activity. *Through the steps of autonomous maintenance, shopfloor employees work with maintenance technicians and engineers toward a common goal—more effective equipment* (see Figure 4-4). By sharing what they know, they can catch many of the problems that cause failures, defects, or accidents.

> ### TAKE FIVE
>
> Take five minutes to think about these questions and to write down your answers:
>
> - Who performs basic cleaning and maintenance on the equipment in your work area?
> - Do you think autonomous maintenance activities would reduce equipment problems in your company? Why or why not?

CHAPTER 4

Figure 4-5 (a). Stabilizing with Autonomous Maintenance

Focused Equipment and Process Improvement

Key Term

Focused equipment and process improvement is the TPM pillar that deals most directly with improving equipment-related losses. If autonomous maintenance and preventive maintenance activities are like exercise and health checkups, focused improvement is like an intense workout tailored to develop strength in specific muscle groups. Autonomous maintenance and planned maintenance improve OEE to a certain level, then help maintain basic operating conditions to stabilize OEE. To raise OEE beyond this stabilized level, companies apply focused improvement (see the left and right sides of Figure 4-5).

Key Point

In contrast to the ongoing activities of autonomous maintenance and planned maintenance, focused improvement involves targeted projects to reduce specific losses. These projects are usually carried out by cross-functional teams that include people with various skills or resources an improvement plan might require. Depending on the target, a focused improvement team may include maintenance technicians, engineers, equipment designers, operators, supervisors, and managers.

44

IMPROVING OEE

Figure 4-5 (b). Targeting Losses with Focused Improvement

Key Point

It's a good idea for companies to attain a basic "fitness" level with autonomous maintenance and planned maintenance before launching focused improvement projects to address specific weaknesses. One reason is to eliminate routine problems (sporadic losses) so you have a clear view of difficult or more significant problems (chronic losses). Another reason is to avoid using a more expensive and time-consuming focused improvement approach for problems that could be addressed through less expensive autonomous maintenance or planned maintenance.

Focused improvement teams use a range of approaches to cut equipment-related losses. They may use 5 Why analysis as a starting point, but there are also approaches that address specific types of losses, such as setup losses and scrap. We will review approaches that deal directly with shortening changeover time and reducing losses from product defects. Finally, we will look at P-M analysis, an advanced version of root cause analysis that is used in focused improvement and quality maintenance.

CHAPTER 4

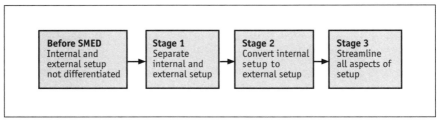

Figure 4-6. The Three Stages of SMED

Quick Changeover

Setup and adjustment time is an improvement target for OEE, since it reduces the time in which the machine is available to make products. Shigeo Shingo, who helped develop key aspects of the Toyota Production System, invented a changeover improvement system called single-minute exchange of die,* or SMED. This system gives a three-stage approach for shortening setup (see Figure 4-6).

Stage 1: Separate Internal and External Setup

In developing SMED, Shingo analyzed changeover operations to determine why they took so long. He recognized that changeover activities could be divided into two types:

Key Terms

- *Internal setup*: setup operations that can be done only with the equipment stopped
- *External setup*: setup operations that can be done while the machine is working.

Key Point

The problem at most companies is that internal and external setup operations are mixed together. This means that things that could be done while the machine is running are not done until the machine is stopped.

---

*Named for the goal of completing changeover within a single-digit number of minutes—9 minutes or fewer.

IMPROVING OEE

Key Point

Example

Stage 1 of Shingo's SMED system involves sorting out the external setup operations so they can be done before the machine is stopped. This alone can reduce setup time by 30 to 50 percent. Typical stage 1 activities include

- Transporting tools and parts to the machine in advance
- Confirming that exchangeable parts are functional before the changeover begins

Stage 2: Convert Internal Setup to External Setup

The next step in the SMED system is to look again at activities done with the machine stopped and find ways to do them while the machine is still active. Typical stage 2 improvements include

Example

- Preparing operating conditions in advance, such as heating a die mold with a preheater instead of using trial shots of hot material
- Standardizing functions such as die height to eliminate the need for adjustments
- Using devices that automatically position the parts without measurement

Stage 3: Streamline All Aspects of Setup

This stage attacks remaining setup time, and includes these approaches to shorten internal setup:

Example

- Using parallel operations (two or more people working together)
- Using quick-release clamps instead of nuts and bolts
- Using numerical settings to eliminate trial-and-error adjustments

---

TAKE FIVE

Take five minutes to think about these questions and to write down your answers:

- How long does a typical changeover take in your work area?
- Can you list the changeover steps that could be performed while the machine is still running?
- Who would you want to have on a setup improvement team, and why?

CHAPTER 4

> Source inspection
> + 100 percent inspection
> + Prompt feedback and action
> + Poka-yoke systems
>
> **Zero Defects**

Figure 4-7. The ZQC System

ZQC (Mistake-Proofing)

The quality rate is an element of OEE. When the equipment that should add value to a product makes a defect instead, it wastes valuable materials and energy—and it can hurt the company's reputation if the defective item reaches a customer. Therefore, quality is an important element of a machine's effectiveness.

Key Point

Many companies think that they are addressing quality issues through inspection that catches defects before they leave the factory. However, *inspection after processing does not eliminate defects, and doesn't necessarily catch them all, either.* Quality cannot be "inspected in." It must be built into the process.

Key Term

Shigeo Shingo carefully analyzed the causes of defects in manufacturing plants and found that random errors were often the most difficult causes to control. To prevent these errors, he developed a mistake-proofing system known as *Zero Quality Control* (ZQC, or "quality control for zero defects").

Key Point

ZQC prevents defects by catching errors and other nonstandard conditions before they actually turn into defects. It ensures zero defects by inspecting for proper processing conditions, for 100 percent of the work, ideally just before an operation is performed. If an error is discovered, the process shuts down and gives immediate feedback with lights, warning sounds, and so on. The basic elements of a ZQC system are summarized in Figure 4-7.

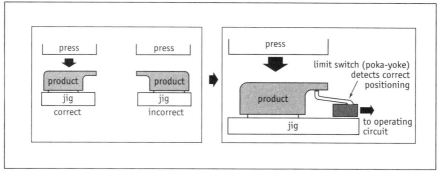

Figure 4-8. A Poka-Yoke Example

Poka-Yoke Systems

Key Point

Because people can make mistakes even in inspection, mistake-proofing relies on sensing mechanisms called poka-yoke systems, which check conditions automatically and signal when problems occur. Poka-yoke devices include electronic sensors such as limit switches and photoelectric eyes, as well as passive devices such as positioning pins that prevent backward insertion of a workpiece. Figure 4-8 shows an example of a limit switch used as a poka-yoke system to prevent processing when the work is placed incorrectly. Poka-yoke systems may use counters to make sure an operation is repeated the correct number of times.

Key Point

The key to effective mistake-proofing is determining when and where defect-causing conditions arise and then figuring out how to detect or prevent these conditions, every time. Shopfloor people have important knowledge and ideas to share for developing and implementing poka-yoke systems that check every item and give immediate feedback on problems.

TAKE FIVE

Take five minutes to think about these questions and to write down your answers:

- What types of actions or conditions can cause defects in your process? At what point could you detect such an action or condition?
- Who would you want on a mistake-proofing team for your process, and why?

CHAPTER 4

Figure 4-9. Chronic Problems Require an Advanced Approach

P-M Analysis

You may have experienced situations when you have to make repeated repairs and adjustments on a recurring problem (Figure 4-9). When a problem comes back, it is usually because the situation is not as simple as we originally thought it was. Our 5 Why analysis may have followed one factor to a deeper cause, but real life is complex and interrelated—several factors often work together to create a particular problem. *P-M analysis is a tool for systematically uncovering and testing all the possible factors that could contribute to a chronic problem such as defects or failure.*

The "P" in P-M analysis stands for "phenomenon"—the abnormal event we want to control. It also stands for "physical"—the perspective we take in viewing the phenomenon. "M" refers to "mechanism" and to the "4Ms"—a framework of causal factors to examine (Machine, Men/Women [operator actions], Material, and Method). P-M analysis is often spelled with a hyphen to distinguish it from abbreviations for preventive or planned maintenance.

The essence of P-M Analysis is to look systematically at every detail so no physical phenomena, underlying condition, or causal factor is missed. Although product defects and equipment failures are the losses most often addressed, P-M analysis can be applied to any loss that involves an equipment abnormality.

P-M analysis involves physically analyzing chronic losses according to the principles and natural laws that govern them. The basic steps of P-M analysis are

How-to Steps

1. *Physically analyzing chronic problems according to the machine's operating principles.* This means understanding—in precise physical terms—what happens when a machine malfunctions. To do this, the team must first understand the physical standard for normal operation.

2. *Defining the essential or constituent conditions underlying the abnormal phenomena.* This means understanding at the physical level what conditions exist when the machine doesn't work right. Examples include the position of the work or the temperature of a cutting tool.

3. *Identifying all factors that contribute to the phenomena in terms of the 4M framework.* This means examining the problem from several viewpoints to uncover factors the team might otherwise overlook.

After going through these steps, the team surveys for the presence of the factors, then tests improvement actions. Figures 4-10 and 4-11 on the following pages are P-M analysis tables. Figure 4-10 shows how information is developed at each step. Figure 4-11 shows how the team checks for factors and tests its improvements.

Key Point

P-M analysis is considered an advanced tool because this level of "detective work" requires more time, resources, and expertise than 5 Why analysis. For these reasons, focused improvement teams may save P-M analysis for complex or costly problems.

TAKE FIVE

Take five minutes to think about these questions and to write down your answers:

- Think of a familiar situation where a machine problem recurs. What do people usually do about it? Can it be resolved with 5 Why analysis, or does it need more analysis?

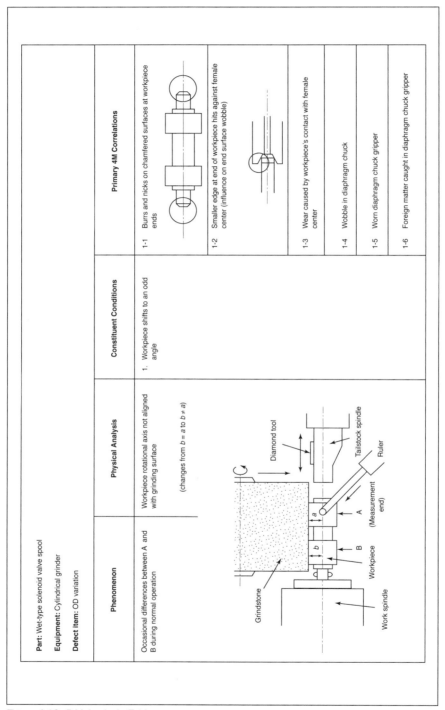

Figure 4-10. P-M Analysis Table

IMPROVING OEE

Part: Wet-type solenoid valve spool
Equipment: Cylindrical grinder
Defect item: OD variation

Check Site	Check Items	Measurement Methods	Survey Findings			Evaluation	Countermeasures	Results
			Standard Values	Measured Values				
Work	Burrs and nicks on workpiece end	Visual	No burrs or nicks allowed	Burrs detected		Action required	1. Revised NC lathe program (previous process) to prevent burrs 2. Added a brush to the top of the grinder chute to remove slight burrs and scale	OK
Work	Taper angle of workpiece end	Industrial microscope	33°	28 to 30 degrees		Action required	Set taper angle to 33° for all workpiece ends so that larger edge OD of workpiece end makes contact with female center	OK
Tailstock	Wear in female center	Dimension gauge	Differential of 0.02	0.03		Action required	Restored (reground)	OK
Tailstock and chuck	Deviation between chuck and center of female center		$\frac{0.002}{50}$	0.002		OK	OK	OK
Chuck	Diaphragm chuck wobble		0.005	0.001		OK	OK	OK
Chuck	Chuck gripper wear	Pin gauge	Within φ 6.6	φ 6.45		OK	OK	OK
Chuck	Dirt and debris caught in chuck gripper	Visual	No debris allowed	Dirt and debris detected		Action required	Attached an automatic air blower	OK

Figure 4-11. P-M Analysis Survey Results

In Conclusion

> **SUMMARY**
>
> - We measure OEE to monitor the condition of the equipment.
>
> - The point of using the OEE measure is to drive improvement. Sustained improvement requires a dedicated approach, with management support.
>
> - When a problem returns, it often turns out that we have been treating the symptoms of the problem, but not dealing with its root cause. 5 Why analysis is a useful tool that brings us closer to the root cause.
>
> - Autonomous maintenance refers to activities carried out by shopfloor teams to help stabilize basic equipment conditions and spot problems early. It changes the old view that operators just run machines and maintenance people just fix them.
>
> - Along with preventive maintenance, autonomous maintenance activities help raise OEE by maintaining proper operating conditions, and stabilize it by detecting abnormalities before they turn into losses.
>
> - Through the steps of autonomous maintenance, shopfloor employees work with maintenance technicians and engineers toward a common goal—more effective equipment.
>
> - Focused equipment and process improvement is the TPM pillar that deals most directly with improving equipment-related losses.
>
> - Autonomous maintenance and preventive maintenance improve OEE to a certain level, then help maintain basic operating conditions to stabilize the OEE. To raise OEE beyond this stabilized level, companies apply focused improvement.
>
> - Focused improvement involves targeted projects to reduce specific losses, carried out by cross-functional teams that include people with skills or resources an improvement plan might require.

- It's good for companies to attain a basic "fitness" level with autonomous maintenance and preventive maintenance before launching focused improvement projects to address specific weaknesses.

- Focused improvement teams use a range of approaches, including approaches that address specific equipment-related losses.

- Setup and adjustment time is an improvement target for OEE, since it reduces the time in which the machine is available to make products.

- Shigeo Shingo invented a changeover improvement system called single-minute exchange of die, or SMED. This system gives a three-stage approach for shortening setup:
    - Stage 1: Separate internal and external setup
    - Stage 2: Convert internal setup to external setup
    - Stage 3: Streamline all aspects of setup

- The quality rate is an element of OEE. Many companies think that they are addressing quality issues through product inspection that catches defects before they leave the factory. In reality, however, this kind of inspection does not eliminate defects.

- To address random errors that cause defects, Shigeo Shingo developed the Zero Quality Control (ZQC) mistake-proofing system.

- ZQC prevents defects by catching errors and nonstandard conditions before they turn into defects. It ensures zero defects by inspecting for proper processing conditions, for 100 percent of the work, ideally just before an operation is performed. If an error is discovered, the process shuts down and gives immediate feedback.

- Mistake-proofing often relies on sensing mechanisms called poka-yoke, which check conditions automatically and signal when problems occur.

- The key to effective mistake-proofing is determining when and where defect-causing conditions arise and then figuring out how to detect or prevent these conditions, every time.

- P-M analysis is a tool for systematically uncovering and testing all the possible factors that could contribute to a chronic problem.

- The "P" in P-M analysis stands for "phenomenon"—the abnormal event we want to control. It also stands for "physical"—the perspective we take in viewing the phenomenon. "M" refers to "mechanism" and to the "4Ms"—a framework of causal factors to examine (Machine, Men/Women [operator actions], Material, and Method).

- The essence of P-M Analysis is to look systematically at every detail so no physical phenomena, underlying condition, or causal factor is missed.

- The basic steps of P-M analysis are

    1. Physically analyzing chronic problems according to the machine's operating principles

    2. Defining the essential or constituent conditions underlying the abnormal phenomena

    3. Identifying all factors that contribute to the phenomena in terms of the 4M framework

- After going through these steps, the team surveys for the presence of these factors, then tests improvement actions.

- P-M analysis is an advanced tool because it requires more time, resources, and expertise than 5 Why analysis.

REFLECTIONS

Now that you have completed this chapter, take five minutes to think about these questions and to write down your answers:

- What did you learn from reading this chapter that stands out as particularly useful or interesting?

- Do you have any questions about the topics presented in this chapter? If so, what are they?

Chapter 5
Reflections and Conclusions

CHAPTER OVERVIEW

Reflecting on What You've Learned

Opportunities for Further Learning

Conclusions

Additional Resources on TPM, OEE, and Equipment-Related Losses
- Training and Consulting
- Packaged Education and Support
- Conferences and Public Events
- Newsletter
- Website

CHAPTER 5

Figure 5-1. Reflect on What You've Learned and What Is Most Useful to You

Reflecting on What You've Learned

Key Point

An important part of learning is reflecting on what you've learned. Without this step, learning can't take place effectively. That's why we've asked you at the end of each chapter to reflect on what you've learned. And now that you've reached the end of the book, we'd like to ask you to reflect on what you've learned from the book as a whole.

Take ten minutes to think about the following questions and to write down your answers.

- What did you learn from reading this book that stands out as particularly useful or interesting?

- What ideas, concepts, and techniques have you learned that will be most useful to you as your company applies overall equipment effectiveness and TPM? How will they be useful?

- What ideas, concepts, and techniques have you learned that will be least useful as you apply OEE and TPM? Why won't they be useful?

- Do you have any questions about OEE or TPM? If so, what are they?

REFLECTIONS AND CONCLUSIONS

Opportunities for Further Learning

Here are some ways to learn more about overall equipment effectiveness and TPM:

How-to Steps

- Find other books, videos, or trainings on this subject. Several are listed on the next pages.

- Investigate software that can help you record OEE data for reporting.

- If your company is already using OEE or implementing TPM, visit other departments or areas to see how they are applying the ideas and approaches you have learned about here.

- Find out how other companies have applied OEE and used it for improvement. You can do this by reading magazines, newsletters, and books that cover OEE, other aspects of TPM, and lean manufacturing, as well as by attending conferences and seminars that share implementation examples and pointers.

Conclusions

Overall equipment effectiveness is more than a set of measurement steps. Used to its potential, it is a fundamental approach for improving the manufacturing process. We hope this book has given you a taste of how and why this approach can be helpful and effective for you in your work.

CHAPTER 5

Additional Resources on TPM, OEE, and Equipment-Related Losses

Training and Consulting

One of the best ways to learn how to use OEE is to apply it under the guidance of an expert. Trainers can bring you through the steps of implementing TPM and applying OEE to your equipment; experienced consultants can help you address specific issues in your plant. Productivity, Inc. provides training and consulting to support team-based TPM implementation.

Packaged Education and Support

Packaged education is not a substitute for training or implementation with an expert, but it can help prepare people for implementation, and give information on specialized topics. Productivity, Inc. offers a wide range of packaged education and support materials related to TPM, OEE, and equipment-related losses.

OEE Software

Arno Koch, *OEE Toolkit: Practical Software for Measuring Overall Equipment Effectiveness* (Productivity, 1999)—A software package originally developed by a TPM consultant and programmer to meet his clients' need for a cost-effective way to capture and report OEE data. It features an easy-to-use interface for configuration and data entry, a wide range of printable color-coded graphs, and a complete user's manual with guidance for defining what to measure. A demonstration CD-ROM is available.

TPM and OEE

Masaji Tajiri and Fumio Gotoh, *Autonomous Maintenance in Seven Steps: Implementing TPM for the Shopfloor* (Productivity, 1999)—The most comprehensive book available in English for planning and managing a complete autonomous maintenance program. (Previously published as *TPM Implementation*.)

Japan Institute of Plant Maintenance, ed., *Autonomous Maintenance for Operators* (Productivity, 1997)—A Shopfloor Series book on key autonomous maintenance activities. Topics include cleaning/inspection, lubrication, containment of contamination, and one-point lessons related to maintenance.

REFLECTIONS AND CONCLUSIONS

Japan Institute of Plant Maintenance, ed., *TPM for Every Operator* (Productivity, 1996) — This Shopfloor Series book introduces basic concepts of TPM, with emphasis on the six big equipment-related losses and OEE, autonomous maintenance activities, and safety.

Kunio Shirose, ed., *TPM Team Guide* (Productivity, 1995) — A Shopfloor Series book that teaches how to lead TPM team activities in the workplace. It includes a section on developing and presenting project reports, and offers guidance with teamwork issues.

Kunio Shirose, *TPM for Workshop Leaders* (Productivity, 1992) — Describes the hands-on leadership issues of TPM implementation for shopfloor TPM group leaders, with case studies and practical examples to help support autonomous maintenance activities.

Charles J. Robinson and Andrew P. Ginder, *Implementing TPM: The North American Experience* (Productivity, 1995) — Describes how TPM fits into an overall manufacturing improvement strategy for Western companies. A real-world perspective on what works and what doesn't, and an educational tool for middle and upper management. Chapter topics include OEE, autonomous maintenance, and implementing TPM in a union environment.

Nachi-Fujikoshi Corporation, ed., *Training for TPM: A Manufacturing Success Story* (Productivity, 1990) — A classic case study of TPM implementation at a world class manufacturer of bearings (a winner of the prestigious PM Prize). This is the complete story of how the company eliminated 90 percent of equipment-related losses in just three years.

Kunio Shirose, et al., *P-M Analysis: An Advanced Step in TPM Implementation* (Productivity, 1995) — Describes an effective step-by-step method for analyzing and eliminating recurring equipment problems caused by multiple or complex factors. This is the best resource in English for this advanced problem-solving approach that was introduced in Chapter 4 of this book.

Tokutaro Suzuki, ed., *TPM in Process Industries* (Productivity, 1994) — Adapts TPM measures and activities for the specific needs of process and large-equipment-based industries.

Tel-A-Train, *TPM* video series (distributed by Productivity) — An informative overview series in four programs: Introduction, Overall Equipment Effectiveness, Preventive Maintenance, and

Predictive Maintenance, plus a Facilitator's Guide (includes TPM Pilot Flowchart and overhead transparencies).

Quick Changeover

Productivity Development Team, *Quick Changeover for Operators* (Productivity, 1996)—A Shopfloor Series book that describes Shingo's three stages of changeover improvement with examples and illustrations.

Shigeo Shingo, *A Revolution in Manufacturing: The SMED System* (Productivity, 1985)—A classic book for managers that tells the story of Shingo's SMED System, describes how to implement it, and provides many changeover improvement examples.

Zero Quality Control and Poka-Yoke (Mistake-Proofing)

Productivity Development Team, *Mistake-Proofing for Operators* (Productivity, 1997)—A Shopfloor Series book that describes the basic theory behind mistake-proofing and introduces *poka-yoke* systems for preventing errors that lead to defects.

Shigeo Shingo, *Zero Quality Control: Source Inspection and the Poka-Yoke System* (Productivity, 1986)—A classic book for managers and engineers describing how Shingo developed his ZQC approach. It includes a detailed introduction to poka-yoke devices and many examples of their application in different situations.

NKS/Factory Magazine, ed., *Poka-Yoke: Improving Product Quality by Preventing Defects* (Productivity, 1988)—An illustrated book that shares 240 poka-yoke examples implemented at different companies to catch errors and prevent defects.

Conferences and Public Events

Listening to implementation stories from other companies is a great way to learn new approaches to common issues. Productivity sponsors an annual TPM conference, forums on TPM for the automotive and process industries, and public training events on OEE and other TPM topics. The 3½ day Maintenance Miracle workshop offers hands-on experience in implementing team-based autonomous maintenance activities on machines at a host plant.

Newsletter

Lean Production Advisor—A Productivity publication sharing the best case studies and product reviews related to implementation of lean thinking and use of specific lean manufacturing approaches such as TPM and OEE.

Website

www.productivityinc.com—The Productivity, Inc. website, with information on a full range of products and services related to TPM and OEE.

About the Productivity Development Team

Since 1979, Productivity, Inc. has been publishing and teaching the world's best methods for achieving manufacturing excellence. At the core of this effort is a team of dedicated product developers, including writers, instructional designers, editors, and producers, as well as content experts with years of experience in the field. Hands-on experience and networking keep the team in touch with changes in manufacturing as well as in knowledge sharing and delivery. Drawing from customer input, the team plans and creates effective vehicles to serve the full spectrum of learning needs in an organization.

About the Shopfloor Series

Put powerful and proven improvement tools in the hands of your entire workforce!

Progressive shopfloor improvement techniques are imperative for manufacturers who want to stay competitive and achieve world class excellence. And it's the comprehensive education of all shopfloor workers that ensures full participation and success when implementing new programs. The Shopfloor Series books make practical information accessible to everyone by presenting major concepts and tools in simple, clear language and at a reading level that has been adjusted for operators by skilled instructional designers. One main idea is presented every two to four pages so that the book can be picked up and put down easily. Each chapter begins with an overview and ends with a summary section. Helpful illustrations are used throughout.

Books currently in the Shopfloor Series include:

5S FOR OPERATORS
5 Pillars of the Visual Workplace
The Productivity Development Team
ISBN 1-56327-123-0 / 133 pages
Order 5SOP-B9001 / $25.00

QUICK CHANGEOVER FOR OPERATORS
The SMED System
The Productivity Development Team
ISBN 1-56327-125-7 / 93 pages
Order QCOOP-B9001 / $25.00

MISTAKE-PROOFING FOR OPERATORS
The Productivity Development Team
ISBN 1-56327-127-3 / 93 pages
Order ZQCOP-B9001 / $25.00

JUST-IN-TIME FOR OPERATORS
The Productivity Development Team
ISBN 1-56327-134-6 / 96 pages
Order JITOP-B9001 / $25.00

TPM FOR EVERY OPERATOR
The Japan Institute of Plant Maintenance
ISBN 1-56327-080-3 / 136 pages
Order TPMEO-B9001 / $25.00

TPM FOR SUPERVISORS
The Productivity Development Team
ISBN 1-56327-161-3 / 96 pages
Order TPMSUP-B9001 / $25.00

TPM TEAM GUIDE
Kunio Shirose
ISBN 1-56327-079-X / 175 pages
Order TGUIDE-B9001 / $25.00

AUTONOMOUS MAINTENANCE
The Japan Institute of Plant Maintenance
ISBN 1-56327-082-x / 138 pages
Order AUTOMOP-B9001 / $25.00

FOCUSED EQUIPMENT IMPROVEMENT FOR TPM TEAMS
The Japan Institute of Plant Maintenance
ISBN 1-56327-081-1 / 144 pages
Order FEIOP-B9001 / $25.00

CELLULAR MANUFACTURING
One-Piece Flow for Workteams
The Productivity Development Team
ISBN 1-56327-213-X / 96 pages
Order CELL-B9001 / $25.00

info@productivitypress.com   www.productivitypress.com   1-888-319-5852